NASA JOHNSON SPACE CENTER ORAL HISTORY PROJECT
ORAL HISTORY TRANSCRIPT

WRIGHT: Today is June 3rd, 2004. This oral history interview with Joe Engle is being conducted in Houston, Texas, for the NASA Johnson Space Center Oral History Project. The interviewer is Rebecca Wright, assisted by Sandra Johnson and Jennifer Ross-Nazzal.

Thanks again for coming in today. Our topics today are your participation in NASA's Shuttle Program. Originally slated to begin in March 1978, the Shuttle missions were delayed somewhat, providing you extra time for training. Tell us about this time and how you were able to separate your time for training for your own mission, for STS-2, as well as being a backup for STS-1.

ENGLE: Well, the delay in the launch had only minor frustrations, really. Of course, everybody wanted to get the bird in the air and show that it would fly okay, but the delays, as you implied, really provided us with more time to prepare for, get ready for contingency situations. Those first few flights, the first flight in particular, and even the second flight, and I'm sure the third and the fourth for Gordo [C. Gordon Fullerton] and the troops, there were so many things that we didn't really know for sure how to train for yet, so we probably overtrained in some areas and overtrained in a lot of areas. But we didn't know that at the time and so we always felt that we could use more time, more additional time, to polish skills, to look at more things, to turn over some more rocks. And every rock we would turn over, there'd always be interesting things under it to look at and to prepare for.

I remember very distinctly not having the impression of idling or spinning our wheels or treading water during those delays. We were engrossed in always new things to look at, a list of things that we could go at. I think we at the time thought, anyway, we had prioritized things. Of course, new things would always pop up and some of them would jump to the top of the priority list.

But we were very, very busy getting ready for things, overpreparing, as I mentioned, I think, in retrospect. The guidance system was very suspect, the inertial platform was suspect in how accurate it would be as far as bringing us back to the intended landing site. So we spent a lot of time looking at dry lakebeds all over the southeast part of the country and practicing approaches into those dry lakebeds, and researching the lakebeds to see what the surface area was actually like, whether it would support the Shuttle, and how long they were, and what the prevailing winds were, and when they were wet. It was interesting. It was fun from a pilot's standpoint, and we did develop a great deal of confidence that we could bring the airplane back; we could bring the *Columbia* back and land her anywhere in the southwest part of the country, if the guidance was not accurate.

WRIGHT: Could you share with us some of the preparations that were made with the flight controllers, how you and [Richard H.] Dick Truly worked with the flight controllers in preparing so that you would know how to communicate on board, and just those types of aspects of that preparation?

ENGLE: Because of the launch delays, we did have additional time to prepare for the flights and we got to work very closely and come to know and have a rapport with the controllers and, in fact, all the people in Mission Control, not just flight control, but [Donald R.] Don Puddy, but all the people that were on the console, all his experts on the various systems. We knew them by voice, when we would hear transmissions. Of course, in the real flight, we only would talk to either the CapCom [Capsule Communicator], sometimes pass information on to Don Puddy, the Flight Director, but normally the CapCom was the direct interface. But we knew the people on the consoles. We got to know pretty much who was on the console by what kind of response or direction we were given for certain simulated failures that we'd had during simulations. And that was good; that was really good. We worked as a very, very close-knit team, almost being able to think and read each other without a whole lot of words said.

WRIGHT: When you were named as Commander of STS-2, there was some time, of course, that lapsed between the time that you were named and to the time that you were able to launch. How did your mission change from the time that you were given that assignment to the time that you actually launched?

ENGLE: Probably the biggest change that occurred was more emphasis on being able to make a repair for a tile that might have come off during flight. As you know, John [W. Young] and Crip [Robert L. Crippen] lost a number of tiles on STS-1. Fortunately, none were in the critical underside, where the maximum heat is. Most of them were on the OMS [Orbital Maneuvering System] pod and on the top of the vehicle. But the inherent cause of those tiles coming loose and separating was not really understood, and on STS-2, we were prepared to at least try to fill some of those voids with RSI [Reusable Surface Insulation], the rubbery material that bonds the tile to the surface itself. So in our training, we began to fold in EVA [Extravehicular Activity] training, using materials and tools to fill in those voids, not a lot unlike what we're preparing to do now, when STS-114 flies and we do return to flight. The tile repair techniques really are fairly similar. The materials and tools have been improved so that there's much more confidence in it being a successful repair.

WRIGHT: What do you remember about your EVA training? Where was it done?

ENGLE: It was done in the old Neutral Buoyancy Facility, which started out as the centrifuge building, the round, circular building, and I think it's attached to Building 5, but I don't remember what the number of it is now. A large pool was built in there in place of the centrifuge, when the value of the centrifuge became pretty much nullified, so all the EVA training was done in that large neutral buoyancy pool. That was the predecessor, of course, to the Sonny Carter Training Facility that we have now. The training was very similar to what's done now as far as inflating the suit and balancing the suit with lead weights so you'd be neutrally buoyant, and then practicing the routines, practicing the EVA procedures that we'd use.

The tasks were not only trying the different proposed ways to access areas on the Shuttle by climbing, straddling, and shuffling out on the boom itself, and then the anomaly of closing and latching the payload bay doors if in fact the payload bay door motors had failed. We practiced manually closing them with a pulley arrangement that pulled the doors closed and cinched them and then a manual latch that we would install up in the corners on the latches.

So we did some very, very elementary EVA training, both Dick and I did, and that probably was the major change from the time we started. I think the other major addition or change as we evolved was to incorporate a series of flight test data maneuvers, inputs, into the vehicle during the entry so that the aerodynamic parameters could be extracted from the data that was being recorded, and we were able to then understand more fully what the margins were of the Space Shuttle as you reenter the atmosphere. On STS-1, understandably, the desire was to fly the most benign entry possible, the least disturbed entry possible, and just keep it right in the groove all the way down, not knowing at all how the vehicle would react or respond to that kind of an entry.

The entry was very, very successful and we were therefore able to convince the management that we could aggressively pursue looking at what the margins were so that in future flights we'd know if we could go to a lower angle of attack for more cross range; for example, if we could maneuver at higher roll rates if we had to in an off-nominal situation. So we were able to perform a whole series of maneuvers, which was very, very rewarding from a pilot's standpoint.

I think another unique thing about what Dick and I were able to start looking at was the transatlantic abort mode of operation, or the maneuver. The abort windows were from the time the solid rocket engines burned out, you could then take over manually and fly what was called a return-to-launch-site abort profile, which was flying outbound for a ways and then turning around backwards, flying backwards, and then accelerating back toward the Cape [Canaveral, Florida] until the fuel had burned out of the tank, and then jettisoning the tank and gliding back into the Cape. And that was a very, very uncomfortable maneuver to fly, a very demanding maneuver to fly, and nobody liked the idea that it would certainly be a last-ditch maneuver.

The other, if you had an engine problem, was abort to orbit, or abort once around, which you had to have almost enough energy, the engines had to burn long enough to get you almost to orbit anyway, and you'd go once around the Earth and then come back in and land. And in between those two, was a significant period of time where if you lost one or two engines, you couldn't make it to orbit and you couldn't make it back to the Cape, so you ended up jettisoning and rolling out and trying to make a skipping reentry maneuver, and keep the Orbiter under control until you could get to glide situation and then bail out after you got down below 20[,000] to 25,000 feet. I can't remember the altitude right now.

We practiced these situations, these emergency scenarios in the simulator, and both John and I were very frustrated that we had the airplane under control and were flying it back, but had to throw it away. John and Crip didn't have time or didn't have the luxury of time to work on and develop any kind of a recovery maneuver and procedure, but Dick and I did, because we had a little extra time as their backup crew and also then preparing for STS-2.

We were able to develop a technique that we could, on the outgoing leg, fly a little more of a depressed trajectory, because we didn't have to retain the ability to fly back to the Cape quite so long, because we could pick up the ability to then just press on with the same amount of energy and make it across the Atlantic and land on the west coast of Africa, or I think at the time we had targeted for Spain, to Madrid or Marone. I can't recall which was our base at that time. But it closed the gap, and we had developed manual procedures that we could do that and flew that.

Subsequent to that, then, the guidance people took those profiles and built them into the auto guidance, and we now have the transatlantic abort option during launch, which not only closes the abort gap and gives you good assurance you're going to get the vehicle back intact, but it also, by not lofting the initial climb-out profile of the Orbiter, it allowed more payload. You needed less energy to fly that kind of a maneuver; it allowed more payload manifesting in the future. So that was a rewarding thing that we did, which was a fascinating thing for pilots to do, to cut and try and put that profile together. Not only fascinating and fun for pilots, but it actually had a practical application downstream.

WRIGHT: Maybe not on a practical note, a couple months after the landing of STS-1, you and Dick Truly accepted a cardboard key to the Space Shuttle *Columbia*. What memories do you have of this?

ENGLE: I remember doing that. I don't remember too much about it. I think the hope was that would be a traditional handover of the vehicle to the next crew. It was a fun thing to do. In fact, I think it was done at a pilots' meeting one time, as I recall. John and Crip handed Dick and I the key, and I think there were so many comments about buying a used car from Crip and John, that it became more of a joke thing than a serious traditional thing, and I don't recall that it really lasted very long. I think it turned from cardboard into plywood, and I don't recall that it was done very long after that.

WRIGHT: A fun moment.

ENGLE: A fun moment. Plus, once we got the next vehicles on line, *Discovery*, *Challenger* on line, why, it lost some significance as well. Besides, you weren't really sure which vehicle you were going to fly after that, so you didn't know who to give the key to. [Laughs]

WRIGHT: And you wanted to make sure you had the right key.

ENGLE: That's right. [Laughs]

WRIGHT: Based on the lessons from STS-1, did you have any concerns, or what were your concerns as you prepared for STS-2?

ENGLE: From STS-1's experiences, I think the tile repair was the primary thing that we probably prepared for as a result of STS-1. The entry maneuvers, we had those in mind prior to the flight, and I think we were even working on the transatlantic abort profile before John's flight got airborne.

But when John and Crip actually lost a number of tiles on their flight, the emphasis then was raised. The focus was narrowed to at least having some kind of technique that was at least a try on fixing the tile.

We always, of course, from the beginning we had the RMS, the remote manipulator system, the arm, manifested on our flight, and that was a major test article and test procedure to perform, to actually take the arm, to de-berth the arm and take it out through maneuvers and attach it to different places in the payload to demonstrate that it would work in zero gravity and work throughout its envelope. Dick became the primary RMS responsible crewmember and did a magnificent job in working with the arm people. Sally [K.] Ride, I believe, was one of the primary RMS support people, and they had worked up a very complete, very extensive test profile to run the arm through, and he was pretty consumed in that during a good bit of our training time.

WRIGHT: STS-2 had a number of firsts, with the RMS being one, and you also mentioned Sally Ride. It was the first time that NASA had a female CapCom. She was also involved as a support crew and helped train you and Dick Truly on being able to take the photos for part of the geological assignments. Tell us about that training and what were some of the techniques that you needed to have to take not just pictures, but good photos that could relay the information back to the scientists.

ENGLE: Anticipating; in other words, knowing ahead of time which orbit a particular site was coming up or an opportunity was coming up, and knowing how many, really, seconds it would be until we would be overhead and in a position to get good photos was probably really the key to that, and I'm sure still is. There are some features on the Earth that stand out, and really stand out especially as you go overhead, things like the Straits of Gibraltar and the Florida Keys and Long Island [New York] and Cape Cod [Massachusetts], features that are very distinctive and very prominent.

But there are two things about space flight that make those a little more difficult to take advantage of. One is, the Earth is often very much covered by clouds, so features that you hope to have and you have programmed to take and in your time line and all, may be covered with clouds as you pass over, so it nullifies the opportunity to do that.

The other thing is, when you look other than straight down, as you start to look out toward the horizon, features blend in or disappear into haze pretty rapidly, so those distinct land-sea interfaces and features on the ground, volcanoes, even, and things like that, they quite often are not really perceptible until they get pretty much underneath you, and they go by pretty much in a hurry as well. Things go by about the same angular rate as they do when you're in an airliner, really, traveling. So you do have time; it's not a snapshot thing, by any means, but you don't have a lot of time as they pass underneath.

So you have to anticipate and have the proper lens on for whatever features you want, what magnification you want, and the settings on the camera at the time. The Hasselblads were manual—the apertures and the shutter speeds were set. So we had a table of land targets, geological targets, Earth observation points, I guess they were called, and each one had its own shutter speed, sun angle, aperture setting, lens that you would use, and being prepared as that photo opportunity came underneath was really the major thing. Very little opportunity to say, "Gee, look. Isn't that great down there? Let's go get a camera and take a picture," because by the time you did, it would be gone. That's one of the things we learned early in the program, was that it really was beneficial to have loaded and ready cameras Velcroed to the windows overhead, so that if you saw something come up like that, you could, in fact, grab a camera and take a picture in a minimum amount of time anyway.

WRIGHT: Were you and the scientists satisfied with the products that you brought back?

ENGLE: Oh yes. At that time, we were still in the "Holy cow. Look at that" phase. Features that were just awesome, that folks had not really seen yet, and not nearly so much into the optical light spectrum of vegetation. The experiment we did have was the telescope. [Office of Space and Terrestrial Application 1 (OSTA-1)]. But it was a side-looking imaging device that could penetrate foliage and features that covered the actual surface of the Earth, and look at and actually penetrate down a ways to get subsurface features. For example, we were able to, with that imaging system, find old riverbeds, river paths, through the Sahara Desert that are covered up by sand, and other features, like sunken ships. The *Graf Spe,* that was scuttled by the Germans in Montevideo, Brazil, we passed over it and were able to image it as we went across. It was a great experiment, a fun experiment.

WRIGHT: With the RMS, you also had a camera. Some of the maneuvers that you were trying, with the different cameras and how to maneuver the RMS and then to turn the cameras, I watched the video and noticed that one of the camera shots was you and Dick Truly waving through the windows, so you were experimenting with those. I also noticed that there was a "Hi, Mom" sign. Did you have anything to do with that as well?

ENGLE: [Laughs] That was Dick's. Yes, that was a great move on his part. He had made up that "Hi, Mom" sign and had it ready. Again, that was those exercises with the camera to see what visibility envelopes were available with the elbow camera and effecter camera. They were worked out by Dick and Sally, and I think, Dick, when he was, I guess, going through the routine must have realized that there was an opportunity to pull a good one like that. That was a great one.

WRIGHT: STS-2 launched seven months after STS-1. It actually had a couple delays. It was supposed to launch in October. How did the delays in the launch affect you and Dick Truly in your schedule of training, and also, how did the launch compare to the simulators?

ENGLE: The delay really was not bothersome at all. I don't think we even among ourselves discussed any negative aspect or any disappointment with delays. Of course, we ready to get airborne at anytime, but we also knew nobody was going to jump in line ahead of us, so it was not a concern at all.

On the contrary, much as in STS-1, but I think even more so in STS-2, it gave us more time to prepare for things that had been identified that could use some more preparation for. The transatlantic abort and the entry flight test maneuvers, we call them PTIs now, programmed test inputs now. They were not programmed, they were manual test inputs at the time, and it gave more time to practice and tune and hone those maneuvers so that the input from the crew, my inputs into the stick were optimized to give the proper response from the vehicle, which would allow the best recovery at that data, those parameters, the primary and secondary parameters, by the aerodynamicists and engineers.

So we didn't really have any dejection over the delays. We used every minute to our benefit on that. And morale was high. We kept a very high morale, because we knew we were going to fly and we knew that we had a neat, challenging mission and we were wanting to do the best we could.

WRIGHT: Were the simulators changed or enhanced after STS-1? Was there feedback from that crew?

ENGLE: Yes, there's feedback after every mission. The feedback is sometimes very transparent or translucent to the crew because of the sophistication of the flight control system. The airplane is designed to try and fly with an optimum response to the pilot input, and sometimes obtaining that response pushes the flight control system itself, the hydraulic system and the electrical feedback system, too, pushes it to its limits. So the hydraulic system may be working very, very hard in trying to satisfy what you're asking from the cockpit, and there are ways that you can filter some inputs, ramp some inputs, to knock the edges off of some inputs, that we learned were, in fact, driving the flight control system to or nearly to its limit. But because it was still getting the right reaction from the airplane, the crew didn't realize it, even though the hydraulic system may have been just having its tongue out, trying to keep up with us.

So there were continual improvements to the flight control system, to the avionics systems, but again, they were pretty much invisible to the crew and they were even pretty much—you asked about the similarity between the trainers and simulators and the STA [Shuttle Training Aircraft] landing trainer airplane, with the vehicle. I think our impression was that the simulators and the STA were very, very close to giving the same response impression that the actual vehicle did, and I think some of that may have been a little masked by the fact that having been in a zero-gravity environment for a duration of time really recalibrates your sensory perceptions, and 1-G [gravity] is not your calibration point anymore; zero gravity is your calibration. So when you get back in the pattern, even 1-G, like we're sitting here now, feels very natural. But when you first come back from orbit, it doesn't. You sense that you're really heavy and being pulled down and something isn't right. And when you have a pilot task on top of that, it kind of masks out your ability to detect the subtle changes in a simulator and an airplane.

WRIGHT: Speaking of airplanes, sixteen years before you launched on STS-2, you were in your [North American] X-15 and touched space for just a brief moment in time, and now you were having an opportunity to go back, but this time for a longer duration. Can you share with us your thoughts as you were sitting there on the launch pad getting ready to return to some place that you'd been longing to go?

ENGLE: Yes, as I recall, the only conscious recollection to the X-15 was that at the end of the flight, we would be going back to Edwards [Air Force Base, California] and landing on the dry lakebed and I think that's where I felt for all of the training and all of the good simulation that we received, that's where I felt the most comfortable, the most at home, going back to Edwards. And at the end of the flight, when we rolled out on final approach going into the dry lakebed, that turned out to really be the case. It was a demanding mission and there were a lot of strange things that went on during our first flight, but when we got back into the landing pattern, it just felt like I was back at Edwards again, ready to land another airplane.

WRIGHT: You launched on STS-2, but just two and a half hours into flight, you had a fuel cell that failed and the mission rules dictated that your flight was going to be reduced from 125 hours to 54. What was your reaction when you heard that your flight was going to be decreased in time?

ENGLE: We were disappointed. As I recall, we kind of tried to hint that we probably didn't need to come back, we still had two fuel cells going, but at the time, it was the correct decision, because there was no really depth of knowledge as to why that fuel cell failed, and there was no way of telling that it was not a generic failure, that the other two might follow, and, of course, without fuel cells, without electricity, the vehicle is not controllable. So we understood and we accepted. We knew the ground rules; we knew the flight rules that dictated that if you lost a fuel cell, that it would be a minimum-time mission.

We had really prepared and trained hard and had a full scenario of objectives that we wanted to complete on the mission. Of course, everybody wants to complete everything. Our first sense of disappointment, really, was one of, gee, now we're not going to be able to do all this stuff, and there's been this big investment in all of these things and we're not going to be able to get the data on them. So that was, I think, our first real disappointment. I don't think we consciously thought, well, we're not going to have five days to look at the Earth. I don't think that really entered our minds right then, because we were more focused on how we are going to get all this stuff done.

WRIGHT: You managed to do it, because the mission was declared as 90 percent complete with its mission objectives.

ENGLE: We were able to do it because we had trained enough to know precisely what all had to be done and we prioritized things as much as we could. When our sleep period came—fortunately, we didn't have TDRSS [Tracking and Data Relay Satellite System] at the time. We only had the ground stations, so we didn't have continuous voice communication with Mission Control and Mission Control didn't have continuous data downlink from the vehicle either, only when we'd fly over the ground stations. So when our sleep cycle was approaching, we did, in fact, power down some of the systems and we did tell Mission Control goodnight, but as soon as we went LOS, loss of signal, from the ground station, then we got busy and scrambled and cranked up the remote manipulator arm and ran through the sequence of tests for the arm, ran through as much of the other data that we could, got as much done as we could during the night. We didn't sleep that night; we stayed up all night. Then the next morning, when the wakeup call came from the ground, why, we tried to pretend like we were sleepy and just waking up.

After the flight, I remember Don Puddy saying, "Well, we knew you guys were awake, because when you'd pass over the ground station, we could see you were drawing more power than you should have been if you were asleep." But that was about the only insight they had into it.

The fact that we were up all night, in addition to not getting sleep, which may or may not have been a good plan, in retrospect, in getting ready to do the de-orbit and landing then the next day, we also had a problem with our water in that the membrane that failed on the one fuel cell allowed excess hydrogen to get into our drinking water supply, so we had very bubbly water available. Whenever we'd go to take a drink, I don't remember the percentage, but a large percentage of the volume was hydrogen bubbles in the water, and they didn't float to the top like bubbles would in a glass here and get rid of themselves, because in zero gravity they don't; they just stay in solution. We had no way to separate those out, so the water that we would drink had an awful lot of hydrogen in it, and once you got that into your system, it's the same way as when you drink a Coke real fast and it's still bubbly; you want to belch and get rid of that gas. That was the natural physiological reaction, but anytime you did that, of course, you would regurgitate water. It wasn't a nice thing, so we didn't drink any water. So we were dehydrated as well; tired and dehydrated when it was time to come back in.

In addition, the winds at Edwards, they were very high a couple of orbits before entry, when we were making the entry preparations, and there was a chance that we would not be able to land at Edwards, but would have to divert. So there was a number of interesting things that contributed to the entry.

WRIGHT: Let's talk about the reentry. It certainly was a momentous time, as you had mentioned earlier, that there were some maneuvers that were performed, twenty-nine from what our research says, and they were performed during Mach 24 to subsonic—it was the fastest procedures that had ever been manually performed. You're the only astronaut to have manually flown the Shuttle in and landed it. Could you give us the rationale of why this needed to be done, and can you walk us through during that reentry and tell us what was going through your mind as you were coming back home to Edwards?

ENGLE: The rationale behind the maneuvers was, as I alluded before, we were very anxious to see how much margin the Shuttle had in the way of stability and control authority, how much muscle the surfaces had at different Mach numbers, hypersonic Mach numbers and angles of attack.

Also, in the event that a de-orbit had to be made on an orbit that had excessive cross range to the landing site, in order to get more cross range rather than S-turn back and forth to deplete energy, the technique was to just leave the vehicle in the bank in one direction and keep flying toward the landing site, off your straight ground track toward your landing site. You could increase that cross-range ability by actually decreasing the angle of attack. It allowed the leading edge of the wing to heat up a bit more and would cut down on the total number of missions that a shuttle could fly, but it would allow you to get that extra performance, that extra range, to make it to the landing site.

How much the leading edge would heat up and just how much more lift-to-drag that would give you, turning ability, cross-range ability, was theoretically known and had some wind-tunnel test data, but the wind tunnels are very susceptible to a lot of variables, Reynolds numbers and scale effects and things like that. So you really want to know for sure what you have in the way of capabilities if you ever have to use them, and that's what our purpose was.

During the entry, I would pulse the vehicle in all three axes—in pitch, a step input, rolls, inputs, and rudder kicks for yaw—to see what the effectiveness of the surfaces were during entry, or effectiveness of the flight control system was during entry, and how quickly the vehicle would damp out after being disturbed.

As far as cross-range or the performance capability, at various Mach numbers, at a couple different Mach numbers, we swept the angle of attack, deliberately pushed the nose over, decreased it, I think, 5 degrees, I believe it was, plus or minus 5 degrees, to see how much more cross-range 35 degrees angle of attack gave us than 40 did, and conversely, went above 40 degrees to 45 degrees to see if we had for some reason wanted to lower the heat on the leading edge of the wings, we could pull up to a higher angle of attack. But that would cost us range, down range and cross range, and just how much that did cost, so that in the future, if that was necessary, the flight planners could then program where the de-orbit burn was. If you didn't have as much range, you could make the burn a little bit later, so that you weren't as far from the landing site as nominally planned.

So getting that data to verify and confirm the capabilities of the vehicle was something that we wanted very much to do and, quite honestly, not everyone at NASA thought it was all that important. There was an element in the engineering community that felt that we could always fly it with the variables and the unknowns just as they were from wind tunnel data and always come down the chute. Then there was the other school, which I will readily admit that I was one of, that felt you just don't know when you may have a payload you weren't able to deploy, so you have maybe the CG [center of gravity] not in the optimum place and you can't do anything about it, and just how much maneuvering will you be able to do with that vehicle in that condition? How much control authority is really out there on the elevons? And how much cross range do you really have if you need to come down on an orbit that is not the one that you really intended to come down on?

So it was something that, like in anything, there was good healthy discussions on and ultimately the data showed that, yes, it was really worthwhile to get and, therefore, those maneuvers that we did on STS-2 were programmed into the automatic flight control system, into the entry flight control system so that subsequent to that, those maneuvers continued to be made and data continued to be gotten, but it was done automatically by the computer.

WRIGHT: In the successful landing that, according to the [NASA] Dryden Flight Research Center [Edwards, California], that was watched by more than 200,000 people. Did you have any idea when you were landing, that there were that many people that were on hand?

ENGLE: No. No, I didn't. In fact, I don't think Dick and I even thought about that. We knew there were a lot of people out for John and Crip's landing, but when we got cut from our five days down to two days, I think we figured, well, nobody's going to be there, because nobody knows about this. I don't think we gave a second thought if anybody would be out there for the landing or not.

If I may back up on the entry, I mentioned that we had a vehicle with a fuel cell that had to be shut down, so we were down to less than optimum amount of electrical power available. Let's see. What all else was going on? The winds were coming up at Edwards. We hadn't had any sleep the night before, and we were dehydrated as could be. And just before we started to prepare for the entry, Dick decided he was not going to take any chances of getting motion sickness on the flight, because the entry was demanding, with all these profiles. Dick had the cue card, and the plan was for him to read off the Mach number and the condition for the maneuver and what the maneuver was going to be, just to remind me of what these twenty-nine maneuvers were, so we did them precisely right on the way in. He had replaced his scopolamine patch and put on a fresh one. The atmosphere was dry in the Orbiter and we both were rubbing our eyes. We weren't aware that the stuff that's in a scopolamine patch dilates your eyes.

So we got in our seats and got strapped in, got ready for entry, and I'd pitched around and was about ready for the first maneuver and said, "Okay, Dick, let me make sure we got the first one right," and I read off the conditions. I didn't hear anything back, and I looked over and Dick had the checklist and he was going back and forth and he said, "Joe Henry, I can't see a damn thing."

So I thought, "This is going to be a pretty good, interesting entry. We got a fuel cell down. We got a broke bird. We got winds coming up at Edwards. We got no sleep. We're thirsty and we're dehydrated, and now my PLT's [Pilot] gone blind." [Laughs]

But back to the landing; the maneuvers were not compromised. Fortunately, Dick was able to read enough of the stuff and I had memorized those maneuvers. That was part of the benefits of the delay of the launch was that it gave us more time to practice, and those maneuvers were intuitive to me at the time. They were just like they were bred into me, which I was glad.

It seemed like everything went in slow motion; it was just waiting and waiting for the next maneuver to get that input in and to see what the response was.

But the landing, when we did get back overhead Edwards and lined up on the runway, as I mentioned before, I think one of the greatest feelings that I've had in the space program since I got here, was rolling out on final and seeing the dry lakebed out there, because I'd spent so much time out there, and I dearly love Edwards and the people out there.

In fact, I recall when Dick and I spent numerous weekends practicing landings at Edwards, I would go down to the flight line and talk with guys that I didn't know at the time, because I'd been gone a while, but I knew their predecessors and people on the maintenance line, and go up to the flight control tower and talk with the people up there, and we would laugh and joke with them. I remember the tower operator said, "Well, give me a call on final. I'll clear you." Of course, that was not a normal thing to do, because we were talking with the CapCom here at a Houston throughout the flight. But I rolled out on final, and it was just kind of an instinctive thing. I called and I said, "Eddy Tower, it's *Columbia* rolling out on high final. I'll call the gear on the flare."

And he popped right back and just very professional voice, said, "Roger, *Columbia*, you're cleared number one. Call your gear." [Laughs]

It caused some folks in Mission Control to ask, "Who was that? What was that other chatter on the channel?" Because nobody else is supposed to be on. But to me it was really a neat thing, really a gratifying thing, and the guys in the tower, Edwards folks, just really loved it, to be part of it.

WRIGHT: What type of review did you personally conduct when you were on the ground, of the Orbiter? Did you take a look at it once you got back?

ENGLE: We did. We sure did. I think every pilot, out of just habit, gets out of his airplane and walks around it to give it a post-flight check, I think. It's really required when you're an operational pilot, and I think you're curious just to make sure that the bird's okay. And of course, after a reentry like that, you're very curious to know what it looks like. You figure it's got to look scorched after an entry like that, with all the heat and the fire that you saw during entry, or the glow from the heat during entry.

Additionally, of course, we were interested at that time to see if the tile were intact, if any tile had come off, chips. We lost a couple of tiles, as I recall, but they were not on the bottom surface, lower surface. They had perfected the bonding on those tiles first, because they were the most critical, and they did a very good job on that. But we walked around, kicked the tires, did the regular pilot thing.

WRIGHT: While you were on flight, you received a phone call from President [Ronald W.] Reagan, who was visiting Mission Control.

ENGLE: Sure did.

WRIGHT: Was that expected? Tell us about that moment.

ENGLE: It was not a total surprise, because I recall we got a call from CapCom saying that he was on site and he would be making a quick call, so it wasn't like the phone ringing and you're picking it up and him saying, "Hi, this is Ronnie." But it was a real honor. I remember wishing that I had had more time to think about the right thing to say or something really prophetic to say, but I didn't. It was just one of those, "Yes sir, things are going great. Thank you for calling." It was a very brief call, but it was a real honor to get to talk to him.

WRIGHT: And after returning back to Houston, you had breakfast with Vice President George [H.W.] Bush.

ENGLE: That's right, yes.

WRIGHT: How did that come about? Did you have a chance to have more of a visit?

ENGLE: We did. We had a good chance to visit. In fact, that was held over at the [Johnson Space Center] Gilruth Center, I remember. President Bush, as you know, was such a personable person; still is. I still just enjoy the heck out of reading things that he's said and quoted, and I've had the occasion to meet him and say hello a number of times since. I think the world of that man. But it was very much of an honor at the time, and we had the breakfast. He's a very friendly, very personable guy. Mrs. [Barbara] Bush, as everyone knows, is just—you wouldn't trade her for anything. [Laughs]

WRIGHT: That kind of kicked off your public relations tour. What all did you have to do as part of post-flight activities?

ENGLE: Of course, I think every crew dreams of going to Europe and going to the Alps and skiing and doing things like that. We had the Canada arm, the RMS on board, built by Spar [Aerospace Ltd.] in Toronto, Canada, and we had a great interface and developed a great relationship with the Canadians in preparation for that flight, both training in Canada and training here and installation of the arm into the bird down at the Cape. So our post-flight tour was to be a tour of all of the provinces bordering the U.S. in Canada.

So just prior to the Christmas-New Year's vacation time frame, we started at the Maritime provinces and started working our way west, and I think we got about to Saskatchewan, then after the break, continued on west through the Rocky Mountains—had wonderful experiences there—and on out to the coast, and thoroughly enjoyed it.

We really enjoyed working with the Canadians. We had a number of fun exchanges with them. I know as the payload bay was being finished out and closed out, the Canadians claimed they were having a little trouble with the thermal protection blanket, the last section of the blanket, which was just prior to the shoulder joint, and no one thought anything about it, because it had to be there, but it wasn't a big deal and they assured us it would be there on time.

So very soon before payload-bay closure, the blanket showed up and the Canadians proudly wrapped it around and it had a big Canadian flag on it. Dick and I decided, "Man, we cannot allow that to happen. We're going to have to outdo them. We've got to have a big American flag somewhere on there." So we went down to the dime store—I forget what the name of it was at the time, but we got, I believe it was, a three-by-five flag. It may not have been quite that big, but we got an American flag anyway, and said, "We've got to put this on the aft bulkhead of the payload bay so when the cameras come on, we'll have the cameras pointed toward that flag and that's the first thing that will be downlinked to the ground."

Of course, it had not been through any of the space qualification for materials or anything of that nature, but at that time other things had priority over that. So we got them to sew some Velcro onto the flag and sew Velcro onto the aft bulkhead, and Velcroed that flag onto the aft bulkhead. So we were able to get a one-upsmanship on the Canadians by showing that flag first. Of course, during all the testing, the cameras were on the arm, so the Canadian flag was showing, but we were at least able to counter that a little bit on that.

The other thing with the Canadians I recall was that they got me really good. All of our clothes were packed in storage containers, and you'd pull them out first day one flight and get clean clothes out, clean socks and underwear and all. They had modified some jockey shorts and replaced mine with these modified Canadian jockey shorts, which the side panels on the jockeys were red and the center section was white, with a big maple leaf on the center. And I've still got those. [Laughs] So they did get the last laugh on me on that one.

WRIGHT: Are there any particular lessons or words to the wise that you passed on to [Jack R.] Lousma and Fullerton?

ENGLE: Yes, I think a few things, rather minor, but one, to be ready for the loud explosion and fireball when the solid rocket boosters were ejected. That was not really simulated very well in the simulator, because I don't think anybody really anticipated it would be quite as impressive a show as that, and I think John and Crip—I don't remember that they mentioned it to us, but that caught our attention, and I think we did pass that on in briefings to the rest of the troops, not just to Jack and Gordo, but to everyone else who was flying downstream.

The other thing was the nose-gear derotation after touchdown, after the main gear touches down. The flight-control system in the Orbiter is a rate command system and that means that the vehicle will respond in pitch rate and roll rate and yaw rate, but pitch rate primarily, only when you ask it to do a pitch rate. And it will try everything it can, if you don't ask for anything in the stick, if you don't come out of detent on the rotational hand controller in pitch, and ask it to do a pitch rate one way or the other, it tries very hard not to. It will do everything it can not to, and that's true even after it touches down.

Most airplanes, when you touch down, you have to keep coming back on the stick a little bit and ease the nose down and keep coming back on the stick, because as you slow down, the dynamic pressure's getting lower, so the force on the surfaces are less. They have less force, less muscle, so you have to keep deflecting the surface more and more in order to control the rate of derotation on nose-gear slapdown.

In the Orbiter, you don't do that. As a matter of fact, if you don't deflect the stick, and the nose starts down, the flight-control system senses a pitch rate with no request from the pilot, so it brings the elevon back up to try to hold that attitude. It's kind of an unnatural thing, plus it happens after you've come through the entry and made the approach, a steep approach, and gotten the bird on the ground and you're kind of in a relaxed mode, "I'm back home safe," and so you're not ready for something new like that. And before you know it, the bird has slowed down, the surface is all the way up, saturated in deflection, trying to keep the nose up, and you really don't want the nose up, because it's going to come down and slap down real hard then.

So John got caught with that on the first flight, and we got caught with that as well in that the nose came down harder than I would have liked to have had it touch down. Not really hard, not particularly hard, because I was aware of it. But we passed that on to the following crews, that you have to make a conscious effort, as soon as the main gear is on the ground, to go forward on the stick, which is kind of an unnatural thing to do, because pilots normally come back on the stick. But you have to initiate the derotation and get the nose started down so that by the time the surface gets almost to full up, why, you're nose is on the ground or nearly on the ground. It's a different technique, is all.

We learned early in the flight, and I think we talked about previously in the Shuttle training aircraft, the large lateral force deflectors on the bottom that we took off the STA because they gave the lateral accelerations for high pitch accelerations when you move the stick quickly back and forth. You tend to do that in fighters and we tend to do that in the T-38s when we're training, like very aggressive roll inputs. They're not really necessary when flying a profile like you fly with the Space Shuttle, which is kind of a gradual, slow, easy pattern, and those quick accelerations are not necessary.

So, just pilot technique can be developed where you make gradual ramp inputs and slowly go into bank maneuvers instead of rapidly, and you avoid those uncomfortable side lurches that you get in the Shuttle. You can do it, but there's just no need to and we can avoid that. So just piloting techniques like that come almost naturally, I think.

WRIGHT: It was about four months after your flight that you went to NASA Headquarters and spent some time up there. Share with us how you were selected to work up at Washington, D.C., for a few months.

ENGLE: General [James A. "Abe"] Abramson, Jim Abramson, was the Director of the Office of Manned Space Flight at the time. I had known and met him, and he asked if I would come up there for a short tour. Actually, he said it was my time in purgatory after having such a good deal, getting to fly the airplane. Jim actually was one of the Air Force MOL [Manned Orbiting Laboratory] pilots, and when NASA picked up the six or seven MOL pilots, they put an arbitrary age limit on who they would take, and Abe was just above that age limit. So he took it very well, but he kidded about that. He was going to get even with everybody, because he didn't get to fly down here.

But he asked me to come up and work with him. That was very early in the Space Shuttle history, in its flight history, and at that time, NASA was very actively pursuing payloads to fill up the payload bay and to generate more flight time and more revenue, and he had asked me to come up and work in that particular capacity, to assess—I think his words were "to give a sniff check" to some of the potential payloads to make sure that they were compatible and applicable to a Shuttle type of operation and deployment and integration into the vehicle.

Then in addition to promotional, it was an educational responsibility to potential customers, not just U.S. customers, international customers. I know we did go to South America and we went to Europe and talked with primarily either national entities, governmental agencies, but also some private entities that had potential communications satellites that were candidates to be flown on the Shuttle. That was supposed to be a four-to-six months tour. I was nine months there.

WRIGHT: You remained on the active flight status with the Astronaut Office and you received another command, STS-51-I. Did you know when you returned to Houston that you were going to have that command? Were you told prior to returning?

ENGLE: No, there was no guarantee of it, and certainly I did not have that assignment when I came back. It was quite some time after I came back before I did get a flight assignment, as a matter of fact.

No, part of the agreement was that the tour at Headquarters would be a temporary thing and I would return to the astronaut corps and get back into the flying chain, but at that time there was no set sequence as to who goes when, and when you would get a flight. It was a matter of coming back and getting into the competitive loop again and waiting for a flight, until those flights that had been assigned were already flown.

WRIGHT: What were some of the duties that you were doing while you were waiting for your next assignment?

ENGLE: I know they involved a good bit of development of flight controls and guidance systems and payload integration systems in the simulators. Working some of the EVA problems; by "working" I mean helping to develop and use the tools on mockups and in the neutral buoyancy tank. But I don't recall anything spectacular that I was assigned to at that time.

WRIGHT: Your second mission was vastly different from the first, with not only span of duration, you were going to be up for seven days, but you also had a crew, and your objectives included launching some satellites. And to even add more challenge to it, as you went through the training process, it changed. Your crew changed, your payloads changed. Can you share with us how all those changes affected your training and how you were able to pull it all together?

ENGLE: The second flight really, on the surface, appeared to be less demanding than STS-2, because we had only a crew of two on STS-2, and one of the lessons we learned from those first four orbital flight tests was that the Shuttle, the Orbiter itself, probably represents more of a workload than should be put onto a crew of two. It's just too demanding as far as configuring all of the systems and switches, circuit breakers. There are over 1,500 switches and circuit breakers that potentially have to be configured during flight, and some of those are in fairly time-critical times. Not only is it a high-task situation from a standpoint of just checklisting and getting all of them configured, but the accessibility. Some of them are on the mid deck and some are on the flight deck, so you're going back and forth and around. Having more people on board really reduces the workload of actually flying the vehicle.

Now, as we progressed into the flights into the Shuttle, of course, the missions became more complex and the payloads became more complex, that compensated to a certain extent. But it still allowed, during the boost phase, to have a full crew of three, with a fourth person on the flight deck helping out as well, to share that workload during launch and during entry and landing. Then once on orbit, the crew could then be assigned separate tasks and separate responsibilities on orbit, and that's exactly what we did on our flight, eventually ended up to be 51-I. And that's what every crew does. The commander assesses what all tasks have to be done, and in some cases, what particular skills some of the crew members have and if they're more applicable to a certain task or experiments to be performed, and assigns those workloads and levels out the workload as much as possible.

The biggest difference between STS-2 and 51-I was the fact that there were more people to help out. There was no problem with the size of the crew or the makeup of the crew as far as compatibility or integration of the crew was concerned. It was very compatible, a very well-integrated crew, even with all the changes.

WRIGHT: It was the twentieth mission, and it launched between rainstorms.

ENGLE: Yes, we did. In fact, we made several launch attempts. On one, we felt that we were really going to go. Everything was counting down smoothly and the flight was called due to weather, and we couldn't see any weather at all out the front windows, which was looking straight up, or out the side. But it turned out that there was a thunderstorm within ten miles, and that was the launch rule at that time, and that you don't launch if there's an active cell within a certain radius of the launch pad.

I remember we got out of the bird and we were very disappointed, because we thought we should have gone. The rain shower never did come over the launch pad. So when we were picked up and joined by Mr. [George W.S.] Abbey and John Young, who was flying the weather chase that day, we started grousing, kind of kidding, but we were saying, "Man, we should have gone today. Why didn't we go to day? We had a perfect day."

And they were somewhat disappointed, too, that we weren't able to launch, and we were told, "Hey, you guys are in the cockpit. We'll make the weather call; you be ready to go fly. When we tell you it's time to go fly, it'll be time and you guys will be ready to go."

On the subsequent attempt, we had rain slickers on when we went from the crew quarters out to the van, and it was raining very hard. When we went from the van to ride the elevator up to the bird and got in the bird, we left our slickers there in the white room, big old yellow rain slickers. We got on board and we really didn't think that there was a prayer of us going to fly that day. The reason that we went out to the bird was that they had one more day delay before they had to detank and that would have been two more days, and the weather forecast was not good for the next day anyway.

So we got in the bird and we strapped in and we started countdown. [James D.A.] Ox van Hoften was in the number four seat, over on the right-hand side aft, and [John M.] Mike Lounge was in the center seat aft, and we were sitting there waiting, and launch control had called several holds. Ox was so big that he hung out over the seats as he sat back, and he was very uncomfortable, and he talked Mike to unstrapping and going down to the mid deck so that he could stretch across both those seats in the back of the flight deck.

We were lying there waiting, and it was raining, and raining fairly good. We got down to five minutes or six minutes, and at that time I can't remember whether it was five or six minutes when we started the APUs [Auxiliary Power Units] after we got the call from launch control to start the APUs. [Richard O.] Dick Covey and I looked at each other kind of incredulously and asked them to repeat. And they said, "Start the APUs. We don't have much time in the window here." So he started going through the procedures to start the APUs, and they make kind of a whining noise as they come up to speed. The rest of the crew was asleep down in the mid deck. I think it was Fish [William F. Fisher] woke up and said, "What's that noise? What's going on?"

We said, "We're cranking APUs. Let's go," or something like that.

Dick was into the second APU, and they looked up and saw the rain coming down and they said, "Yeah, sure, we're not going anywhere today. Why you starting APUs?"

We didn't have time to explain to them, because the sequence gets pretty rushed then. So we yelled to them, "Damn it, we're going. We're going to launch. Get back in your seats and get strapped in."

They woke up Ox and Mike, and they got back in their seats, and they had to strap themselves in. Normally you have a crew strap you in; they had to strap each other in. And Dick and I were busy getting systems up to speed and running, and all we could hear was Mike and Ox back there yelling at each other to, "Get that strap for me. Where's my com lead?"

"Get it yourself. I can't find mine."

And they were trying to strap themselves in, and we were counting down to launch. They really didn't believe we were going to launch because it was, in fact, raining, but they counted right down to the launch and we did go. It went right through, a light rain, but it was raining.

Then after we landed and we asked them about that. "Boy, you know, we launched right through rain. I didn't think we were going to be able to launch."

Of course, the response was, "We were flying weather out at the SLF [Shuttle Landing Facility]. Why didn't you tell us it was raining?"

We used the rationale then. We said, "Our job is to be ready to fly. You guys tell us when the weather's okay."

WRIGHT: What a way to go.

ENGLE: Yes, it was a great launch.

33

WRIGHT: And it's such a busy and full schedule that you had, with so much to do on this mission. On the first day you deployed two satellites. How did that impact the Orbiter?

ENGLE: It really didn't. The reason for launching the two on the first day, rather than one on day one and one on day two, was that in opening the payload bay and in cycling the sunshields on the two satellites, the procedures were out of sequence and the remote arm had not been rolled back. The brackets that hold the arm in place had not been rolled back to provide clearance for that sunshield to come back, so when the sunshield was retracted, it hit the arm and deflected it somewhat.

Mike had to then unberth the arm, take the arm and actually push the sunshield back the rest of the way so that the satellite was exposed, and not knowing whether there would be a problem getting it deployed or not. It was deployed early and then we had time to deploy the second one as well, so we deployed the first two satellites the first day.

Then the Syncom [IV-3 Satellite], the Leasat or Syncom, as it was called, either one, that we carried was deployed then the following day, so we actually got a little ahead on the time line, which was good, because the arm had suffered a malfunction as well in that it was not working. One of the joints had to be operated in the manual mode rather than the automatic mode, so Mike couldn't fly the arm like you normally do, moving it around from one place to another. He had to move it, operating an electrical switch, selecting a joint, and then going plus or minus with the electric motor and moving one joint at a time to get to the right positions to use it. That became a concern as to how much that was going to slow us down on the grapple and the capture and then the redeployment of the failed Syncom that we were going to go repair.

WRIGHT: Speaking of that satellite, one of your tasks was to rendezvous with this rather large satellite that had been deployed earlier that year and not working. Share with us your approach.

ENGLE: As I mentioned earlier, the tasks for the flight, one of my objectives was to try and distribute the tasks among all the crew as evenly as possible and to give everybody something really interesting and meaningful that they could come back and talk about. We had three satellites, so the three mission specialists each were given the responsibility to launch one of those satellites. Ox and Fish were designated as the two who would be going EVA to repair the satellite, so Mike was the remote arm operator, and all three of them therefore played a most important and demanding role in the capture and the repair and the redeployment of this satellite.

Dick and I got to ride in the front seats, and I was going to get to land the vehicle, so I gave Dick the responsibility of doing the rendezvous of the Syncom-IV-3, so he did that. He trained in the simulators and practiced in the simulators and did an absolutely superb and excellent job of that rendezvous. My job, really, was to stay out of his way or help him, assist him in any way I could, but to stay out of his way while he flew the rendezvous, until we were actually within about a thousand feet of the satellite. Then he would turn control of the vehicle over to me for the final approach and getting Ox in position to manually grapple it.

When he had completed the rendezvous maneuver and had stabilized and, I think, said words to the effect, "Okay, boss, it's all yours," and I looked up out of the pipper, the crosshairs of the pipper, I kind of expected to see it somewhere in the field of view in the window, but he had flown that rendezvous and perfectly nailed it so that the satellite was right behind the pippers, and I had to look a couple of times, in fact, to see where it was, because it was right in the center of the pipper.

The rendezvous was a beautifully, perfectly flown rendezvous and made the rest of the task very easy. I say "very easy." It didn't mean that there weren't traps that we had to be careful of in the final approach. Flying up to that satellite, you transition from looking out the overhead and flying a control-system orientation, where up fires fore and aft thrusters and down fires the same, and then left and right is pretty straightforward, but roll is different, so you change axes systems.

As you go from the overhead window and it comes down into the aft window view, you change the orientation of the vehicle, and you also change a perspective, where lifting and pushing and pulling take on different axes as you go from one window to the other. In that transition, my first input was one using the old system, so it was the wrong input, and the satellite starting drifting toward the window instead of down into the payload bay, so I reversed controls to get it back out of the way so we wouldn't impact it, and had to kind of restart and bring it back down into the payload bay again. A good lesson learned and one to bring back and warn the next guys to be careful of that.

WRIGHT: What other dangers or situations were you trying to avoid as the two astronauts were doing the EVAs during the two days? What were some of the things that you were doing to protect the Orbiter?

ENGLE: On the initial grappling of the Orbiter, Ox was out on the end of the arm, in a foot restraint on the end of the arm, and we had planned to capture it with a tool that we developed in the three months that we had to get ready for it. It was like a big towel rod that had clips on the end. The Syncom was supposed to be rolling very slowly in a roll, and the two trunion joints, which are like ball hitches on a pickup truck to pull a trailer with, there are four of them, two on each side. They're used to mount the satellite on to the rails in the payload bay of the Shuttle. We had developed this tool that as two of those came around, as two of those trunions or balls came around, Ox could clip this rod onto them and secure it and stop it and then have a handle to move it around with, because the satellite had not been built to be handled at all on orbit.

When we arrived and saw it first out the window, we saw and found that it was not spinning in a slow, stabilized spin maneuver, but actually the rotation had been slowed down by [Henry W.] Hank Hartsfield's [Jr.] crew's attempt to jar the latch to bring it to life, to make the micro switch contact to bring it to life, and that rotation had been slowed down to a certain extent. But the theory was that the four months that the satellite had spent rotating around the Earth and going through the Earth's flux field took some more rotational energy out of it to the point where it was not stable anymore. It was just randomly kind of tumbling in space. As we approached the satellite, the grappling tool, it was obvious, was not going to be of any use at all and Ox was going to have to just grab it by his hands.

It was a 15,000-pound satellite and about fifteen feet in diameter, which was a huge satellite, and the separation joint surface, we were warned, might have sharp edges where the pyrotechnics separated it from its booster, from its shroud, and we were told, whatever you do, don't get close to that. Well, when we got up to it, it turned out that that was the only really good place that Ox had to grab the satellite, and he looked carefully to make sure there weren't any sharp edges. But he grabbed it by that edge of the sunshield to stop it and to slowly get it rotated around where he could grab theses trunion bolts and stop them right in front of him, and then take the tool and actually mash it on manually. It was a tremendous job of adapting and doing some real-time adaptation on orbit, EVA, on Ox's part, to capture that satellite and to get it so that he could have a hold of it.

The piloting task during that time was keeping Ox in position to do this all the time, by flying the Orbiter and keeping him positioned, because Mike Lounge on the arm really was restricted in how helpful he could be, since he had to fly the arm with single joint. He couldn't automatically fly Ox around and keep him in position.

We accommodated for those things and captured the vehicle, and Ox held it and tried to position the Orbiter to move it down where Fish on the other side could put another handling bar on, and then Ox could then put the grapple fixture on the outside. Then Fish held it while Mike brought Ox back into the payload bay. We took the foot restraint off and Mike then went back up and grabbed the satellite with the arm.

From then on, it actually was pretty straightforward. It took a little more time to orient and position the Syncom, presenting the surfaces to Ox and Fish, where they could remove panels, hotwire the avionics to the battery bus with a wiring harness that we had taken up, and essentially make the electrical repair to the satellite.

Then the redeploy, again, was not as taxing, not as stressful, really, as the capture and getting it secured, but it was a matter of positioning Ox again on the arm and letting him give some pushes on the one towel rod that he had left on. It started spinning, so to get it stabilized so we could then back off and then they could fire the engine to take it on up to geosynchronous orbit. And every time he would push, the satellite, of course, would not only take on rotational, but also translational energy, and it would start to move away. It was a matter of flying Ox back up and keeping him close in so that when the towel rod came by again, why, he could grab it again and give it another push and spin it a little bit faster. It took about three or four of those shoves, as I recall.

WRIGHT: Quite a sight, I would imagine, to watch it.

ENGLE: It was a phenomenal sight. You bet. You bet it was. As a matter of fact, as it was spinning and going away, we flew the Shuttle down around to line Ox up with the satellite so that we could get a picture of him and the satellite in the background. He gave it a salute and then the Charles Atlas sign.

WRIGHT: Speaking of pictures, your crew also did some Earth observations. Any expertise you were able to share with them on how the best way of taking those photos to bring back the shots that the scientists were looking for?

ENGLE: Well, not me personally, other than the continual warning to be ready ahead of time, because you won't recognize your target until it's time to take the picture, then you've got to have it or it's too late. But by that time, a great deal of Earth observation photography had been done and our instructors were very, very good in setting up the charts that would show the times, the orbit, the times, and all the camera settings that were necessary to do it, and what kind of features were desired and what kind of targets of opportunity to really concentrate on, too, to be ready for. By that time it was standard practice to have cameras Velcroed all around the windows of the glass boat to be ready.

WRIGHT: Were you able to see more this trip, since it lasted a little bit longer than your first one?

ENGLE: Oh yes. Yes, you bet. There was more time to look out the window, and that, I think, was one of the neatest experiences about space flight, having the time to look back at the Earth. The professional gratification that comes, for a pilot anyway, comes with the landing, the entry and the landing and touching down and rolling out. I think the most memorable thing is, when you really don't have a meaningful task to do at all, is to look out the window and look back at Earth. It's a very, very inspiring experience to see how thin, how delicate the atmosphere is, how beautiful the Earth is, really, what a beautiful piece of work it is. And to see the features go by.

Sultan [Salman Abdulaziz] Al-Saud was assigned to our crew initially, when one of our payload satellites was the ARABSAT. He was assigned as a mission specialist on our crew, and when he eventually did fly, I think he said it better than anybody has. He said, "The first day or two in space, we were looking for our countries. Then the next day or two, we were looking at our continents. By about the fourth or fifth day, we were all looking at our world."

Boy, it's one of those things that I said, "God, I wish I'd have thought of that. I wish I'd have said that," because I thought that was classic. It really was.

WRIGHT: Tell us about your crew, how you were able to work so well with your crew. Since you had so many changes and mixing and matching, how did it all come together and what was special about them?

ENGLE: Well, the crew, as you mentioned, changed continuously from our first assigned mission until the final eventual one. In fact, our last change came within the last three or four months of flight, when we took on the added task of rendezvousing and repairing Syncom-3. We had to drop payload specialists, because we needed that payload and that room in the cabin for EVA equipment and for tools.

The crews rotated fast enough there was never time to develop any problems or concerns. The folks who were assigned temporarily to the crew, the eventual five of us that went—we all had a total attitude of concentrating on what needed o be done and getting that done, and as we got closer into flight, particularly with the repair aspect of the Syncom satellite, we all were so busy that we didn't have time to develop any kind of irritants or things that bothered each other. We just all seemed to be working together. Everybody had a job that they had to do and they'd show up at the simulator prepared for that session, and that was kind of the story of the crew the whole way through, just a totally dedicated, totally prepared, very competent crew. A lot of talent on that crew.

WRIGHT: Your second flight and also your second Orbiter. Were there differences in flying *Discovery* compared to *Columbia*?

WRIGHT: There weren't any at all that I could perceive. I know folks have asked that, and quite frankly, from an airplane-handling qualities standpoint, I was very, very pushed to find any difference between *Enterprise* and the two orbital vehicles, *Columbia* and *Discovery*, other than the fact that *Enterprise* was much lighter weight and, therefore, performance-wise you had to fly a steeper profile and the air speed bled off quicker in the approach and landing.

But as far as the response of the vehicle, the airplane was optimized to respond to what pilots tend to like in the way of vehicle response. So when you made an input on the stick, it was really transparent what was happening inside the systems and inside the airplane itself. It tried its very hardest to give the optimum response asked, and we really didn't change those responses very much.

There were some things that would have been nice to have had different on the Orbiter, and still are, and that is the hand controller itself. It's not optimized for landing a vehicle. It really is a derivative of the Apollo rotational hand controller, which was designed for and optimized for operation in space, and since that's where the Shuttle lives most of the time, it leans toward optimizing space operations, rendezvous and docking and those types of maneuvers, where you have a definite breakout force in the hand controller from the detent. The stick force gradient, how much harder you have to push for additional deflection, is not as important and as apparent as it is on an airplane. In fact, some of the airplane characteristics are not optimized for space, so it's a compromise stick and a compromise controller.

WRIGHT: After STS-4, the ejector seats were taken out. Did that give you any cause of concern or any cause of thought?

ENGLE: No, no. The ejection seats probably were a comfort to a certain extent in the approach and landing tests, because they could be used pretty much anywhere in the profile, even while you were mated to the [Boeing] 747. If there was a problem, if something drastic happened, why, you could bail out at any time. But on *Columbia,* I don't know. I guess they were there primarily so if there was a gross malfunction in the guidance system and you entered and came subsonic over the middle of a forested area or a mountainous area and there's no place to land, you could at least get into a stabilized glide and bail out. But they really didn't do any good at all on launch, certainly not on orbit.

The idea of the ejection seats, the envelope that they provided you an escape capability was very, very small. They did take up a lot of room and a lot of weight, and they did limit the crew to two people, which, as I mentioned earlier, was a real workload for the crew, a real heavy task load for the crew.

WRIGHT: Are there any other thoughts or memories you'd like to share with us about the second mission or anything connected with it before we move on? How was the landing? You didn't have so many maneuvers on this landing as you did the first one.

ENGLE: No, the maneuvers were all automatic on this one. It was just a matter of as you came up to a Mach number and an altitude or a condition, the maneuver was there ready to happen, and you could inhibit it by going to the keyboard and inhibiting that maneuver so it wouldn't do it, but, no, the entry was much more relaxed.

Having been in orbit for even a short period of time, like we were on STS-2, and experienced an entry and a landing and all, there was much less anticipation. You had a feel for what was going to happen and I was ahead of the game more, I know, on the second flight than on the first flight. The nose gear touchdown was not nearly as hard on the second one. We were fortunate; we had good landings on both flights, so there was a very smooth touchdown on both flights, on the lakebed again, which I was glad of, because it was at Edwards again. And I know it was much more practical to recover the Shuttle at [NASA] Kennedy [Space Center, Florida] than at Edwards. It saved having to haul it all the way across country, but there was just something about that lakebed out there at Edwards and the environment that I really loved landing out there.

[End of interview]

www.ingramcontent.com/pod-product-compliance
Lightning Source LLC
Chambersburg PA
CBHW081800170526
45167CB00008B/3262